NATURE'S BEAUTIFUL ORDER

An Introduction to the Study of Animals Taught by the Classical Naturalists

Student Guide

Christopher O. Blum & John A. Cuddeback

MEMORIA PRESS

AF785045

MEMORIA PRESS
www.MemoriaPress.com

NATURE'S BEAUTIFUL ORDER
An Introduction to the Study of Animals
Taught by the Classical Naturalists

STUDENT GUIDE
Christopher O. Blum & John A. Cuddeback
Contributing Author: Anna McCain

ISBN 978-1-5477-0099-8

Second Edition © 2018 Memoria Press | 1021

Cover illustration by Albert Bierstadt

All rights reserved. No part of this book may be reproduced in any form
by any means without written permission from the publisher.

CONTENTS

CHAPTER 1: The Cat

READING NOTES

1. **Longitudinal** – In anatomy, this term refers to something extending from the head toward the tailbone. A lateral line would be perpendicular, extending across the width rather than the length.
2. **Transverse** – There are multiple body planes that are used to describe how anatomical parts are located in relation to each other. One such plane is the transverse plane. The transverse plane can be thought of as a flat plane, separating the body in half, perpendicular to the spine in an animal.
3. **Quadruped** – An animal that walks on four feet.
4. **Organic and Inorganic** – An organic substance contains the element carbon and is often associated with things that exhibit characteristics of life. An inorganic substance does not contain carbon. When talking about the living world, most substances will be organic, while inorganic substances are more commonly studied when examining the nonliving thing.
5. **Nitrogenous** – Something that is nitrogenous contains the element nitrogen. In animals, when describing nitrogenous tissues, this implies that the animal forms amino acids, necessary for building protein in the body. The cat is used for this example, while plants are mentioned as being non-nitrogenous. This is reasonable, as plants do not produce protein the way that a cat or other animal does.
6. **Carbonic Acid** – Carbonic acid is a molecule composed of hydrogen, carbon, and oxygen that plants use to obtain the carbon they need to be organic (i.e., living) things.

COMPREHENSION QUESTIONS

1. Why is the cat valued as a domestic animal?

____

____

2. What is the Latin word from which the name "cat" derives?

____

____

____

3. Why does it seem likely that the cat would have been first domesticated in Egypt?

____

____

____

____

____

____

4. How is the domestic cat chiefly different from the wild cat?

____

____

5. What is the normal life span of a domestic cat?

6. How many kittens might the owner of a mated pair of cats expect to have every year?

7. Why is knowing the history of the domestic cat and its various breeds not "scientific knowledge" in the most strict sense?

8. Why is it important to recall that we naturally learn by making distinctions?

9. What does it mean to say that an organism is a "complex whole in which all the parts are reciprocally ends and means"?

10. Which four chemical elements are the principal ones into which living things can be dissolved after death?

11. What does it mean to say that an organism is "made up" of these elements?

12. Should an organism be understood to be a collection or mixture of these elements?

13. How do living things differ from nonliving things?

14. What are the most basic activities shared by all animals—that is, those activities that indicate they are both alive and animals?

15. How can we negatively characterize the difference between all animals and plants?

16. How are the higher animals, such as the cat, further differentiated from plants?

17. What, then, is an animal?

CHAPTER 2: The Lobster

READING NOTES

1. **Vertebrate vs. Invertebrate** – A vertebrate animal has a spine, while an invertebrate does not. Animals are often split into these two categories when determining their characteristics.
2. **Regenerate** – Regeneration is the process of regrowth of something that has been removed. Certain simpler animals exhibit massive regenerations, such as that of a worm cut in half that can regenerate its other half. Regeneration is far more rare in complex animals. For example, in humans, the liver is the only organ that will actually regenerate. The converse of regeneration is repair, in which something does not fully regrow, but functionality is restored through the growth of scar tissue.
3. **Nervous System** – The nervous system includes the parts of an animal that send signals to and from all different parts of the body to coordinate movement. Simply put, it is composed of the brain, spinal cord, and nerves.
4. **Sternum** – A long, flat bone located in the center of the chest.
5. **Abdomen** – The abdomen of an animal is often called the belly and contains most of the digestive organs of the animal.
6. **Transverse Joint** – A joint that lies along the transverse plane in the body.
7. **Segmentation** – The separation of individual segments of a body. In crustaceans, the separation of each of the three body regions into smaller segments.
8. **Cephalo-thorax** – The thorax is the part of an animal's body between the head and the abdomen. In lobsters, the head is fused to the thorax, creating the cephalo-thorax.
9. **Dorsal vs. Ventral** – Dorsal and ventral describe two different areas of the surface of an animal's body. The ventral side of an animal is the front side, or the lower side in quadrupeds, usually the side with the front of the belly of the animal. The dorsal side is the back side, usually the side with the back of the animal.
10. **Anterior vs. Posterior** – The anterior is the front of the animal, and the posterior is the back. For a human, ventral is the same as anterior, and dorsal is the same as posterior. For quadrupeds, however, the anterior side is the side toward which the animal's eyes are facing, while the posterior is the opposite side. Therefore, the two sets of terms describe different reference points in the animal.
11. **Somites** – In crustaceans, individual body segments found in the abdomen. These include the exoskeleton of the crustacean.
12. **Appendage** – A part of an invertebrate that extends out from the body.

COMPREHENSION QUESTIONS

1. What is the essential task of the student of natural history?

____

____

2. What is meant by the word "order" when applied to the great chain of being?

____

____

____

3. What order does Aristotle follow in his treatise of natural history, the *History of Animals*?

4. What is Aristotle's method in describing the parts of animals?

5. Why does Aristotle begin and end *History of Animals* by considering the animals as wholes?

6. Why is it particularly difficult for us to understand the invertebrates?

7. In what way are the lower animals less intelligible in themselves?

8. Why is it ultimately unsatisfactory to call a group of animals by the name "the invertebrates"?

9. What characteristic did Georges Cuvier propose as the most likely clue to the natural groupings of the animals without backbones?

10. What are the names of Cuvier's four "branches" of the animal kingdom?

11. Why was Cuvier's choice of the nervous system as the key to the lower animals appropriate?

12. What does a first glance at the lobster show us?

13. What do we learn from a closer look at the lobster's abdomen?

14. What evidence does Mivart bring forward in support of the conclusion that the cephalo-thorax (head and chest region) of the lobster essentially consists of the same segments as the abdomen?

15. How does the nervous system of the lobster differ from the human nervous system?

16. What is the arrangement of limbs in the class Articulata (the articulates), such as is revealed in the lobster?

17. What lesson about the great chain of being is first learned by considering the lobster as an example of the class Articulata (the articulates)?

CHAPTER 3: The Cuttlefish

READING NOTES

1. **Ganglion** – A number of gathered nerve cells used to send nervous signals instead of a brain and spinal cord.
2. **Radiating** – Referring to radial symmetry, symmetry in a circular shape around one central axis. (Think: starfish.)
3. **Circulatory System (Circulation/Circulating)** – The system of organs in an animal used to transport blood throughout the body to deliver oxygen.
4. **Digestive System (Digestion)** – The system of organs in an animal used to digest nutrients and excrete waste out of the body.
5. **Generative Organs (Reproductive System)** – The system of organs in an animal used to reproduce (i.e., create a new animal of the same kind).
6. **Toothed Horny Margin** – This phrase is referring to the edges of each of the little suckers on the arms of a cuttlefish. The rim of each of the acetabula is rough and ridged with little "teeth."
7. **Lens of the Eye** – The lens is located behind the iris and pupil of the eye. It is a thin, transparent layer that helps to focus light taken in by the eye.
8. **Taxonomic Class** – Taxonomy is the practice of identifying different organisms. Organisms can be classified according to many different subgroupings: Domain, Kingdom, Phylum, Class, Order, Family, Genus, and Species. *Nature's Beautiful Order* deals with the Kingdom Animalia, containing many different phyla of animals. Each of these phyla can be separated into many classes, and so forth.
9. **Respiratory System** – Respiratory actions are those in which an animal breathes and is therefore able to oxygenate its blood.
10. **Egress vs. Ingress** – Egress is output, and ingress is input.
11. **Locomotion** – The ability to move from one place to another.
12. **Calcareous** – Containing calcium carbonate, a chalky substance that is used as a source of calcium.
13. **Cartilaginous** – Made of cartilage, a firm but flexible tissue found in multiple body structures in animals.
14. **Salivary Glands** – A gland is an organ that is used for the secretion of substances. The salivary gland is particularly used for the secretion of saliva.
15. **Gullet** – The passage that food takes from the mouth to the stomach.
16. **Crop** – A muscular pouch on the gullet that is used for storing food and pre-digesting it before the food moves to the stomach. Humans do not have a crop.
17. **Vein vs. Artery** – Veins and arteries are used to transport blood throughout the body. Arteries carry oxygen-rich blood away from the heart out to the body. Veins carry oxygen-deficient blood away from the body toward the heart to be sent to the lungs for re-oxygenation.
18. **Bifurcate** – Divided into two branches.
19. **Dilate** – When something widens or opens.
20. **Contractile Sac** – An organ that is used to pump fluid in the body.
21. **Cilia** – A short, hairlike structure that vibrates and can aid in movement.
22. **Auricles** – A small opening resembling an ear.
23. **Testis vs. Ovary** – The testis and ovary are used in reproduction. The testis are found in males and are where the seed is produced. The ovary is found in females and are where the egg is produced. Reproduction will be covered a little more thoroughly later in this book.

24. **Viscid** – Thick and sticky.
25. **Cerebral** – The primary part of the brain or ganglial structure to do with complex sensory and motor functions.
26. **Pedal** – Relating to the feet.
27. **Cerebral Ganglia** – A pair of nerve bundles found at the anterior end of some animals.
28. **Contraction (Contractile)** – The tensing or tightening of something; anatomically, usually a muscle.
29. **Muscular Fibers** – The individual strands of proteins composing the muscles.
30. **Alimentary Canal** – The entire chain of organs by which food moves from the mouth to the anus.
31. **Chitinous (Chitin)** – A hard, fibrous substance that forms the exoskeletons of certain animals.
32. **Auriculo-Ventricular** – Auriculo refers to the atria of the heart, small contractile vessels. Ventricular refers to the ventricles of the heart. This adjective just describes a heart that has both of these features.

COMPREHENSION QUESTIONS

1. How did Aristotle separate the mollusks from the other invertebrates?

____

____

2. Are clams and oysters mollusks even though they have hard shells?

____

____

____

____

3. If we take the cuttlefish as our typical mollusk and consider it from the outside as a whole creature, what do we see as its defining feature?

____

____

____

____

4. What are the tentacles of the cuttlefish ordinarily used for?

____

____

5. Do cuttlefish tentacles have another use?

____

____

____

____

6. What is the mantle of the cuttlefish?

7. How does the cuttlefish move through the water?

8. How is the cuttlefish able to move in one direction and feed from the opposite one?

9. What is the cuttlebone?

10. What lessons may we learn from Dr. Mivart's misunderstanding of the cuttlebone?

11. How is the cuttlefish nourished?

12. How does the cuttlefish digest hard and bony prey?

13. How does the circulatory system of the cuttlefish aid it in its search for prey?

14. How is the nervous system of the cuttlefish organized?

15. What resources does the cuttlefish make use of against potential enemies?

16. How is the cuttlefish different from the lobster?

17. How is the cuttlefish different from man?

CHAPTER 4: The Sea Urchin

READING NOTES

1. **Venom** – A substance secreted by animals that is poisonous, usually injected through biting or stinging.
2. **Spheroidal** – Three-dimensionally spherical or elliptical in shape.
3. **Calcified Plates** – A flat plate that has hardened over time due to calcium compounds being converted into calcium carbonate.
4. **Superior vs. Inferior** – Anatomical terms of reference. Superior refers to something that is above, and inferior is below. For a human, something superior will be closer to the top of the head, and inferior will be closer to the bottom of the feet.
5. **Anus** – The opening through which solid waste leaves the body.
6. **Genital Duct** – An anatomical tube through which reproduction is able to occur.
7. **Vertical Tract** – Vertical simply means running vertically in the body; a tract is a system of organs connected to perform one overarching function (i.e., the digestive tract).
8. **Apical (Apice)** – This term is usually applied to sponges and animals that have a base attached to a surface. The apical end of the animal is opposite the attached end. In the case of the sea urchin, the apical end is opposite to that of the oral end, containing the mouth.
9. **Oral** – Oral refers to the mouth of an animal.
10. **Articulation** – A joint, where two bones are attached for the purpose of motion.
11. **Ligament** – A short band of tough, fibrous, flexible tissue that connects bones or cartilage or holds together a joint.
12. **Tubercle** – A small rounded protuberance from the body.
13. **Convex vs. Concave** – Convex refers to a surface that curves outward, while a concave surface curves inward.
14. ***Ligamentum teres*** – The *ligamentum teres* is a ligament situated in the hip between the pelvic bone and the top of the leg, attaching the two but allowing for movement and rotation.
15. **Rosette** – A rose-shaped structure.
16. **Alveoli** – A bony socket for the root of a tooth.
17. **Mesenteric Filaments** – Cords of tissue that are connected between the mesentery, a tissue lining the walls of the abdomen in humans and around the location of the digestive tract in sea urchins, and the intestines.
18. **Eye-spot** – A spot found on some invertebrates that is pigmented and light-sensitive, though not nearly as complex as the human eye.
19. **Apertures** – An opening or hole.
20. **Auditory Organ** – An auditory organ is involved in hearing.
21. **Visceral Cleft** – An opening in the embryonic development of an animal along the esophagus.
22. **Nidus** – An origin site for development or growth.

COMPREHENSION QUESTIONS

1. Where did Aristotle meet the lowly sea urchin?

2. Where does the name urchin come from?

3. What does a sea urchin without its spines look like?

4. What does a closer look at the structure of the sea urchin reveal?

5. What are the two kinds of bony plates?

6. How are the plates of the urchin's skeleton organized?

7. How might we best picture the organization of the ambulacral and interambulacral plates?

8. How does the sea urchin move?

9. What is Aristotle's lantern?

10. What does the sea urchin eat?

11. How is the nervous system of the sea urchin organized?

12. In what characteristics does the sea urchin most notably differ from the lobster?

13. In what characteristics does the sea urchin most notably differ from the cuttlefish?

14. Are all of the lower invertebrates built on the same general radial and pentagonal plan as the sea urchin?

15. Has Cuvier's fourth branch of the animal kingdom—the Radiata—stood the test of time?

16. Is there anything that can be learned from Cuvier's notion that the Radiata or Zoophytes belong together in one group?

17. What general lesson is to be gained from the consideration of these simplest of animals?

CHAPTER 5: The Bee

READING NOTES

1. **Thorax** – The part of an animal's body between the head and the abdomen.
2. **Solitary vs. Gregarious** – A solitary animal doesn't live in a herd with others, but prefers to be alone. A gregarious animal lives in a group with other animals and is considered to be social.
3. **Entomologist** – Entomology is the study of insects and how they interact with the natural world.
4. **Metamorphosis** – Metamorphosis occurs when an animal, usually an insect, undergoes a transformation from a juvenile (larval) stage to its adult form.

COMPREHENSION QUESTIONS

1. How did Aristotle divide animals according to their mode of life?

____

____

2. What makes a gregarious animal to be properly a social animal?

____

____

3. What is the first way in which we can tell that the honeybee is a social animal?

____

____

____

4. What are the three types of honeybees?

____

____

____

5. Is the queen bee the ruler of the hive?

____

____

____

____

6. Describe the composition of the beehive.

____

____

____

____

7. Is the hexagonal design of the honeycomb significant?

8. What does the honeybee harvest from flowers?

9. What is honey?

10. How are new bees born?

11. What is the difference between the worker bee and the queen bee?

12. What is a drone?

13. How does a colony of bees reproduce itself?

14. How can a swarm of bees be domesticated?

15. What technique helps the beekeeper handle his bees?

16. What does it mean to say that the bee is a social animal?

17. What lesson may we take from the bee as a social animal?

CHAPTER 6: The Trout

READING NOTES

1. **Operculum** – A structure that covers up an aperture.
2. **Lamella** – A thin layer, membrane, or tissue.
3. **Capillaries** – Sections of fine, branching blood vessels.
4. **Aorta** – The main artery of an animal's body.
5. **Branchial Fibers** – The fibers that form the gills of a fish.
6. **Desiccation** – A state of extreme dryness.
7. **Auricle vs. Ventricle** – Both are openings allowing blood flow into and out of the heart. Auricles are located at the top of the heart and bring blood in, while ventricles are located at the bottom of the heart and pump blood out.
8. **Compression vs. Dilation** – Compression describes the shrinking or collapsing of a structure, while dilation is the opening or widening of a structure.
9. **Migratory** – A migratory animal migrates throughout the year to retain stable living conditions (i.e., birds flying south for the winter).
10. **Fly-fishing** – A type of fishing where the "fly" on the end of the line mimics a fly landing on the surface of water, which fish will eat for food.

COMPREHENSION QUESTIONS

1. What did Aristotle counsel us to admire in the more humble animals?

2. Why do we designate the vertebrates as the "higher" animals?

3. What is the organ that makes a fish to be a fish?

4. What is the task of the gills?

5. How does the structure of the gills make this function possible?

6. Where are the gills located on the body of a fish?

7. How do the gills come in contact with fresh (that is, oxygenated) water?

8. Why does a fish out of water die?

9. Why are some fish able to breathe on land?

10. How does the heart of a fish function?

11. What organs in the fish correspond to the limbs of the other vertebrates?

12. How is the fish propelled through the water?

13. What sign does Cuvier offer in support of his contention that "the entire structure of the fish is adapted for swimming"?

14. How is the fish able to eat without limbs?

15. What sort of knowledge of nature does a good fisherman require?

16. How do fly fishermen catch their prey?

17. What general lesson is to be gained from the consideration of the fly fisherman's art?

CHAPTER 7: The Frog

READING NOTES

1. **Tympanum** – The eardrum or tympanic membrane, used for hearing.
2. **Palate** – The roof of the mouth.
3. **Juxtaposed** – Set directly across from one another.
4. **Mandibles** – The lower jaw or jawbone.
5. **Pelvis** – The bony structure at the base of the spine where the legs attach.
6. **Glutinous** – Gluelike and sticky in texture.
7. **Visceral Arches** – Pieces of cartilage or bone that help attach the jaws with the skull and help support the throat.
8. **Plumose** – Possessing multiple fine, branching filaments.
9. **Filaments** – A slender, threadlike structure.

COMPREHENSION QUESTIONS

1. What are the two names of the class of creatures to which the frog belongs?

2. What characteristics do the amphibians share with the reptiles?

3. What characteristics do the amphibians share with the fish?

4. What is the first characteristic of the amphibians that we would be likely to notice if we were to hold a frog in our hands?

5. If we were to attempt to describe the frog's appearance, what features would most easily distinguish it from its brother amphibian the salamander?

6. What is the peculiar function of the frog's skin?

7. What is peculiar about the frog's tongue?

8. What do the limbs of the amphibians have in common with those of the higher vertebrates?

9. What is peculiar about the structure of the frog's pelvis?

10. What do frogs do in the winter?

11. How is the tadpole different from the adult frog?

12. What are the main features of the tadpole's development that cause it to be termed a metamorphosis?

13. How do the amphibians differ from the fish?

14. How do the amphibians differ from the reptiles?

15. How do the amphibians differ from birds?

16. What surprising development led Dr. Mivart to see in the metamorphosis of the amphibians a possible model for the emergence of the different kinds of living things?

17. Why should discussions about "evolution" come after a general introduction to living things?

CHAPTER 8: The Turtle

READING NOTES

1. **Temperature Regulation** – The process by which an animal maintains a stable internal body temperature. These methods will differ based on whether the animal is warm-blooded or cold-blooded.
2. **Pulmonary Cells** – Cells that make up the lungs of an animal.
3. **Asphyxiation** – The state of being deprived of oxygen, leading to suffocation.
4. **Carapace** – The hard upper part of the shell of a turtle.
5. **Plastron** – The lower part of the shell of a turtle.
6. **Vertebral Column** – Another name for the spine of an animal, which is composed of vertebrae.
7. **Pericardium** – The multilayered membrane that encloses the heart.

COMPREHENSION QUESTIONS

1. What is the sensible characteristic that the cold-blooded vertebrates present to us?

____

____

____

2. What is another way of explaining the difference between the cold-blooded and warm-blooded vertebrates?

____

____

____

____

____

____

3. Why may the warm-blooded vertebrates be called higher than the cold-blooded ones?

____

____

____

____

____

____

4. What are the leading characteristics common to all the reptiles?

____

____

____

5. How did the reptiles get their name?

6. What is the structure of the reptilian (and amphibian) heart?

7. What is the defect of the three-chambered heart compared to the four-chambered heart?

8. How do the lungs of reptiles differ from those of mammals and birds?

9. How are turtles distinguished from the other reptiles?

10. Which of the turtle's bones form part of its carapace?

11. What are the three turtle habitats identified by Fabre?

12. What does Fabre mean when he says that the turtle is an animal turned inside out?

13. Why is Darwin's *The Voyage of the Beagle* a valuable text for the natural historian?

14. What is most striking about the Galapagos tortoise?

15. How does the Galapagos turtle thrive in such a hot, dry place?

16. What is most striking about the reproductive habits of the Galapagos turtle?

17. How is the Galapagos turtle a good example of the class of animals known as the reptiles?

CHAPTER 9: The Canada Goose

COMPREHENSION QUESTIONS

1. At what time of year does the Canada Goose leave the south for its breeding grounds to the north?

2. What does Audubon think is the reason why Canada Geese form pair bonds prior to northward migration?

3. What do the mated pairs of geese do immediately after their arrival in the north?

4. What part of the male goose's body does he use as his chief weapon in his struggle to win the female mate of his choice?

5. What is the role of the gander after the eggs have been laid?

6. Where do geese place their nests?

7. What sorts of things does the Canada Goose eat?

8. Who are some of the natural predators of the Canada Goose?

9. How long does it take the goslings to learn how to fly?

10. When do the geese leave the north?

11. What shape does the migrating flock of geese take?

12. What sort of winter abode do the geese seek?

13. Is Audubon correct to ascribe emotions to geese?

14. Why do birds migrate?

15. How do we know that the cost of migration is high?

16. Is migration worth its high cost?

17. How does the annual migration of birds show forth the order of nature?

CHAPTER 10: The Miracle of Flight

READING NOTES

1. **Musculature** – The arrangement of muscles in an animal.

COMPREHENSION QUESTIONS

1. What did Aristotle take the defining characteristic of the birds to be?

2. What would need to be added to Aristotle's account to make it complete?

3. What functions do the feathers serve?

4. What internal features of a bird help it to fly?

5. How is the bird's musculature particularly apt for flying?

6. What is Aristotle's teaching about the habits of animals?

7. What sense do birds chiefly rely upon to find their food?

8. What is the characteristic flight of the "little gray birds," the Chickadee and the Titmouse?

9. How do the Nuthatch and the Goldfinch fly?

10. What sound does the Goldfinch make at the bottom of each segment of its curved flight?

11. What is the origin of the expression "as the crow flies"?

12. How do flocks of Blackbirds and Starlings fly?

13. Why does the Tree Swallow need to be such an acrobatic flier?

14. When and where are you most likely to spy Blue Jays at their most active?

15. Why are the ravens, owls, hawks, and eagles such commanding fliers?

16. What does the Red-Tailed Hawk cry out in its flight high above the ground?

17. Why do different birds have different habits of flight?

CHAPTER 11: Day's Herald Bird

READING NOTES

1. **Larynx** – Another term for the voice box, the larynx is an organ at the top of the neck that holds the vocal cords and forms an air passage to the lungs.
2. **Trachea** – Another term for the windpipe, the trachea is a long tube reinforced with rings of cartilage that conveys air to the lungs.
3. **Oviparous Vertebrata** – Vertebrata is the class of vertebrates. Oviparous means that they reproduce by laying eggs.
4. **Forelimbs** – The limbs found toward the anterior end of a quadruped.
5. **Gizzard** – The muscular part of a bird's stomach used for grinding up food.

COMPREHENSION QUESTIONS

1. What familiar bird sings "cheer-il-ee, cheer-up, chee-eer-io" from the top of trees and fence posts early on spring mornings?

____

____

2. What is voice?

____

____

3. What is the organ of the voice in man?

____

____

4. How is the human larynx constructed?

____

____

____

____

5. How is human vocal sound produced?

____

____

____

6. What do the ventricles and superior ligaments of the larynx do?

____

____

____

7. What is the difference between voice and words (i.e., between vocal sound and language)?

8. Why do birds lack soft mouths?

9. Why are most birds unable to form their vocal sounds into words?

10. What organ is responsible for the bird's voice?

11. Do songbirds have a more impressive syrinx than other birds?

12. Why are the voices of songbirds so loud, while the birds themselves are so small?

13. How do the calls of birds assist them in the task of nourishing themselves and their offspring?

14. Why is there a great chorus of birdsong every spring?

15. What sorts of human behavior can have a positive effect upon the populations of songbirds?

16. What is remarkable about the song of the Mockingbird?

17. Why might the Wood Thrush sing so late into the summer?

CHAPTER 12: Birds at Home

COMPREHENSION QUESTIONS

1. What is the greatest subject that nature presents to us for our contemplation?

2. What is the science of ecology?

3. What fascinated Aristotle about the nests of birds?

4. Why was Audubon puzzled by the nesting habits of the Song Sparrow?

5. What does Audubon's passage about the nesting of Song Sparrows teach us about the attitude the good naturalist takes to the study of nature?

6. How is Audubon's passage on the Song Sparrow illustrative of the proper first principle of biological inquiry?

7. What answer to Audubon's question is suggested by the observations of Margaret Morse Nice?

8. What puzzle remained to Mrs. Nice after her observations?

9. Which birds often forage together in mixed flocks in the autumn and winter?

10. How is the Blue Jay beneficial to his fellow birds?

11. How is the Blue Jay nasty toward his fellow birds?

12. How does the Yellow Warbler defend her brood against the Brown-Headed Cowbird?

13. How are crows beneficial to farmers?

14. How can farmers encourage birds to harvest insects from their fields?

15. What sort of antics is the Chickadee known for?

16. Why was the Wood Thrush the "feathered favorite" of John James Audubon?

17. May it be reasonable to see birds as signs of the angels?

CHAPTER 13: The Groundhog

READING NOTES

1. **Viviparity** – The quality of reproducing by giving birth to live young.
2. **Ectothermic** – Gives off heat.
3. **Endothermic** – Takes in heat.
4. **Dentition** – The arrangement and condition of teeth in an animal.
5. **Mastication** – Chewing, grinding up food before digesting.
6. **Setae** – A stiff, hairlike structure.
7. **Torpid** – Mentally and/or physically inactive, lethargic.

COMPREHENSION QUESTIONS

1. Why is the study of mammals generally more interesting and important than the study of other animals?

2. What are the four things, used here in our consideration, that characterize mammals?

3. What insight into mammalian life can we have from the characteristic of being warm-blooded?

4. What special place does hair have in mammalian life?

5. What is unique about the dentition of mammals?

6. What does the dentition of mammals especially reveal?

7. What important aspect of mammalian life is caused by being viviparous?

8. How are rodents distinct from mammals?

9. What is rodent dentition, and what can we learn from it?

10. What is significant about dentition in animals for our study of natural history?

11. How do the limbs of a groundhog compare to human limbs?

12. Where do groundhogs live?

13. How is the groundhog's body well-suited to where it lives?

14. Are groundhogs edible?

15. What is hibernation?

16. Why does Audubon praise God the Creator after a consideration of the hibernation of groundhogs?

17. When do groundhogs have their young?

CHAPTER 14: The White-Tailed Deer

READING NOTES

1. **Rumen** – The first stomach of a ruminant, used to partially digest food or cud.
2. **Lachrymal** – Related to crying and tears.
3. **Conception** – Also known as fertilization, the process by which development of new life begins.

COMPREHENSION QUESTIONS

1. Does "ruminants" name an order of mammals?

____

____

____

____

____

____

2. How does the ruminants' dentition reflect their eating habits?

____

____

____

____

3. What especially sets ruminants apart from other herbivorous mammals?

____

____

____

____

4. Where does the name ruminant come from?

____

____

____

____

____

5. Why are ruminants especially important for the human diet?

____

____

____

6. What characteristics of white-tailed deer tend to excite wonder in those who observe it?

7. In what ways have deer been used to serve the needs of men?

8. Why does Audubon say that deer cause but little injury to the fields of the planter?

9. Of what does the diet of deer normally consist?

10. What can be said of does as mothers?

11. Do deer have keenly perceptive senses?

12. What is the end or goal of reproduction in animals?

13. How do mammals reproduce?

14. Is nursing of young considered part of reproduction?

15. Why is timing so important in reproduction?

16. What is the rutting season?

17. Why is there male competition in rutting season?

CHAPTER 15: The Carnivora

READING NOTES

1. **Rectum** – The final section of the large intestine, leading to the anus.
2. **Predation** – The preying of one animal on another.
3. **Paroxysms of Asthma** – A sudden attack of being unable to breathe.
4. **Alluvium (Alluvial)** – Fertile soil left by flowing rivers.

COMPREHENSION QUESTIONS

1. Does "Carnivora" name an order of mammals?

2. Are Carnivora the only carnivores?

3. Do Carnivora only eat meat?

4. How are Carnivora well-suited to eat meat?

5. How do Carnivora serve humans?

6. How can a person improve his powers of observation of the natural world?

7. Why do we consider the skunk's organ of spraying exceptional?

8. How does the skunk spray its enemies?

9. How effective a means of self-defense is the skunk's spray?

10. How successful a predator is the skunk?

11. Is the skunk useful for humans?

12. Is the black bear completely black?

13. Is the black bear a carnivore?

14. Does the black bear prefer meat or plant matter?

15. Are black bears considered edible?

16. Are black bears dangerous?

17. Do black bears hibernate?

CHAPTER 16: Farm Friends

READING NOTES

1. **Protein** – Found in many foods, abundantly in meat, proteins function to build and repair muscles and tissues in the body.
2. **Lacteal** – Relating to milk.
3. **Bacteria** – Small microorganisms that feed on different compounds to grow and reproduce.

COMPREHENSION QUESTIONS

1. How is our contemporary understanding of farming different from that of earlier times?

2. Is it a matter for concern when so many people live in isolation from farm life?

3. What are the two things with which we associate cattle on the farm?

4. Who is the "saint of the barnyard"?

5. Does a milk cow care who milks it?

6. Why, according to Mr. Cobbett, is it good for children to learn to milk a cow or to perform other such barnyard labor?

7. Are pigs in the same "order" as cows?

8. What difference does not being ruminants make for pigs' place in the farm?

9. What basis is there for asserting that pigs can produce more meat than any other farm animal?

10. Why can pigs be such an important part of the family farm economy?

11. Why is the fact that pork makes excellent salted meat so important?

12. What two parts of the pig yield meat that is the object of so much delight and praise?

13. How can ham be compared to wine?

14. What is the pig's temperament?

CHAPTER 17: Man the Upright Animal

READING NOTES

1. **Bimanous** – Having two hands.
2. **Biped** – Having two feet.
3. **Median Line** – The middle line of a human, extending from the head through the center of the body.
4. **Scapulae** – Anatomical term for the shoulder blade.
5. **Girth** – The measurement around a person's waist.
6. **Cervical Ligament** – A ligament found in the neck.
7. **Clavicle** – Another name for the collarbone.
8. **Dexterity** – Ability to perform fine motor skills, usually particularly describes hand movement.

COMPREHENSION QUESTIONS

1. Why is the study of human anatomy so important?

____

____

____

2. What feature of the human anatomy most of all separates man from other animals?

____

____

____

3. What are some features of man's anatomy that make it clear that being upright is his most natural posture?

____

____

____

____

____

____

____

4. How are the limbs of man different from the limbs of animals such as monkeys and their cousins?

____

____

____

____

____

5. Why is his upright posture so significant?

6. To what does Aristotle connect the upright posture of man?

7. What further point does St. Thomas Aquinas make about the fittingness of man's upright posture?

8. How is the freedom of man's hands connected to the arts?

9. What is an art?

10. Do lower animals have arts?

11. Has nature provided man with what he needs for warmth, self-defense, and hunting?

12. How do men and lower animals use their senses differently?

13. What greater significance does man's use of his senses for observation have?

14. What does St. Boethius say about man's stature?

CHAPTER 18: Man the Steward

COMPREHENSION QUESTIONS

1. Does the natural world around us belong to man?

2. What is a steward?

3. What does it mean to say that man is a steward of the natural world?

4. How can we know that man should be a steward of the natural world?

5. How is human work connected to stewardship of the natural world?

6. How can agriculture be a good example of stewardship?

7. What attitude is the foundation of true stewardship of the natural world?

8. Can everyone exercise stewardship of the natural world?